AUG 2 7 1992

Where Animals Live

The World of a Falcon

Adapted from Mike Birkhead's *The Falcon over the Town*

**Words by
Virginia Harrison**

**Photographs by
Mike Birkhead**

Oxford Scientific Films

Gareth Stevens Children's Books
MILWAUKEE

Contents

The Kestrel, a Kind of Falcon, and Where It Lives	3
The Kestrel in Cities and Towns	4
The Kestrel's Body	6
The Kestrel's Head	8
Masters of the Air	10
Food and Feeding	12
The Hunt	14
Preparing to Breed	16
Egg-laying	18
Growing Up	20
The Home Life of Kestrels	22
Friends and Neighbors	24
Other Aerial Predators	26
Living with People	28
Life in the Town	30
Index and New Words about Kestrels	32

Note: The use of a capital letter for a kestrel's name means that it is a *species* of kestrel (for example, American Kestrel). The use of a lower case, or small, letter means that it is a member of a larger group of birds.

The Kestrel, a Kind of Falcon, and Where It Lives

Kestrels belong to the family of birds called falcons. They are hunting birds, or birds of *prey*, with sharp, curved beaks, good eyes, and *talons* on their feet. The kestrel is smaller than the average bird of prey, about as big as a crow. Kestrels are at the top of the "food chain," with smaller animals below them. They also dominate their *habitat*, like the tiger in the jungle.

Kestrels can be found all over the world. There are 13 different kestrel *species*. The most common are the American Kestrel of North America and the Common Kestrel of Europe, shown here in London.

Kestrels can live in many different habitats. Some live in valleys, some in mountains. They can even be seen in the busiest cities (above).

The Kestrel in Cities and Towns

Although most kestrels do not feel at home in an area with humans, the American and Common Kestrels do. They have learned to perch on a chunk of metal as well as natural cliffs and old fence posts.

Of course, *urban* kestrels don't look any different from those that live in the fields. But they do behave differently.

Their habitats are very different, and they have to learn to adjust to the new ways of hunting and nesting, and the different kinds of foods. City kestrels must also become used to being around humans.

The Kestrel's Body

Both the European (above) and American Kestrels have reddish-brown *plumage*. Their two strong yellow feet have three toes in front and one in back.

The European Kestrel is a little over 1 foot (30.5 cm) long, with a wing span of about 2.5 feet (70-80 cm). The smaller American Kestrel is about 8 inches (20 cm) long, with a wing span of 1 foot 9 inches (50-60 cm). The male kestrel (below) has more colorful plumage than the female (above).

American Kestrels (above) have much more detailed plumage than the other kestrels, and they are readily identifiable. Young and female kestrels are usually brown with less striking markings.

Although the kestrel is a large bird, like all birds, it has hollow bones. Kestrels are lighter than they look.

A good sign that kestrels live nearby is a feather on a branch or on the ground.

7

The Kestrel's Head

Kestrels have the typical head of all birds of prey. They have large round eyes and sharp beaks. Under their feathers on their heads, they have very powerful ears.

Kestrels' eyesight is superb. They can see three times as well as we can. They can see lots of detail and judge distance well. When flying at such high speed, this is a plus.

Kestrels can see a mouse in a field from their hovering position of 20-30 feet (6-9 m) above.

Kestrels appear never to close their eyes, but they do. They have a clear piece of skin that closes over the eye when they are attacking prey. It helps moisten the eye, too.

Their curved, sharp beaks can be used to kill prey, eat prey (like our teeth), and something else unique to birds. Birds clean their feathers by *preening* them with their beaks.

Kestrels' nostrils are not used for smelling, but for breathing. They have very small tongues, too, but they are too small to have taste buds.

Masters of the Air

The kestrel is streamlined and has special flight feathers to give it lift. It is able to take off, hover, and soar. Its feathers are very strong.

All falcons are fast, streamlined flyers. Peregrine Falcons are the fastest. They reach speeds of 155 mph (250 kph) when *stooping* on their prey. Kestrels (above) are much slower. They can reach over 62 mph (100 kph). But usually they fly at around only 20 mph (32 kph).

The kestrel can stay motionless in the air by beating its wings rapidly and pointing itself into a strong wind. This is called "hovering." When hovering, a kestrel fans out its tail feathers as well as its wings.

Birds have hollow bones that make them light. Birds also have *air sacs* inside to help keep them in the air. They can stay aloft on air currents without using their wings. This is called "soaring."

When a kestrel hovers, it keeps its head still and watches for prey below. Not all birds can hover.

Food and Feeding

Another sign of kestrels in the area is their pellets, or balls of fur and bone they cannot digest.

Kestrels eat meat. They are called "carnivores." Depending on what is in their habitat, they may eat small mammals or birds, or frogs or beetles. In towns, kestrels often feed on rodents or sparrows. Most of what they eat is torn up first by their sharp bills.

Kestrels often choose a certain post or tree to go to with their captured prey. Their pellets are a good sign of where they may be *roosting*. And if you soak a pellet in warm water, you may be able to tell what the kestrel has been eating!

The Hunt

Kestrels hunt by day on their own. They find their prey either by flying and searching or by perching and watching. This bird or its chicks (below) are easy prey for a kestrel.

Once the kestrel spots its prey, it swoops down and catches it. The kestrel may eat the large prey right away, or carry the lighter prey away. Kestrels can eat the smallest prey while flying.

Kestrels often attack their prey directly. The small bird sitting on this ledge (above and below) is easily trapped by the fierce talons.

Kestrels may take food from other birds, and their food may in turn be stolen from them. They will even scavenge for food or *carrion* along the road.

Urban kestrels are unafraid of attacking their prey while humans are close by. In fact, they often hunt for small birds in parks.

Preparing to Breed

Before breeding season in the spring, kestrels perform courtship rituals to find a mate. Each also starts to form a *territory* to defend.

During breeding season, the kestrels perform aerial displays, trying to attract a mate. They call to each other with their "kee, kee, kee," or "killy, killy, killy." They even battle over territory, and sometimes they hurt each other.

During their aerial displays, both the male and female circle and chase each other.

It is important for the female to choose a strong partner that will be able to provide food for her before and during nesting, and for their offspring. The male feeds the female as a part of courtship, too.

Kestrels are monogamous. That means they choose one mate each season and stay together until the end of the season.

Egg-laying

This picture of European Kestrel eggs shows how little material kestrels need for a nest. They often nest on window ledges, but the most natural place for them is in an old tree that has a hole in it.

Because life is easier for urban kestrels, they often lay eggs sooner than the country kestrels. While the female *incubates* the eggs, the male hunts for food.

Both the male and female kestrel have a *"brood patch"* on their breasts that has a rich supply of blood. The patch is warm and aids in the incubation of the eggs.

It takes about four weeks for the chicks to grow inside the eggs. An "egg tooth" helps them crack the egg open. When they are older, they stretch their wings to get ready to fly (above).

Growing Up

The chicks are pink and practically featherless when they first hatch. They are blind, too. As they grow, their nest becomes more and more cramped. The parents are very careful to avoid showing *predators* where the nest is. They never fly directly to the nest.

The strongest chicks push the weaker ones aside in the struggle to eat the food the mother brings. Often the weakest die, but it assures that the offspring will be strong.

As the chicks grow, they develop adult feathers. Before then, they have *downy* white feathers and need their mother to keep them warm.

Most young kestrels are ready to leave the nest when they are about 30 days old. They become more and more active as they prepare for their first flight. When they have their flight feathers, they have *fledged*.

These young kestrels were hatched in a ventilation shaft.

The Home Life of Kestrels

The newly fledged kestrels stay in their parents' territory for about a month. Before they set off on their own, they "play games" with each other. They may chase after a crow or bathe in a puddle. When it comes to food, they do compete with each other.

Some kestrels do not *migrate* very far from their parents. Others do. It depends on the climate. In North America, kestrels generally migrate south for the winter. Some places are well-known to bird watchers for the many birds that migrate there.

It is exciting to see an American Kestrel soaring through the air, high above the ground. ➡

Friends and Neighbors

As predators, kestrels are avoided by many other animals. They do share urban settings with other animals, however. Raccoons like the same places that kestrels like. And in England, it is no longer uncommon to find foxes at dumps and in urban areas. Like the kestrel, it has adapted to the human environment, and it thrives there.

↑

The most common urban mammal is the rat. The Brown Rat (above) and the Black Rat (below) thrive on human garbage. Both the kestrel and the fox prey on rats.

↓

Other Aerial Predators

The only other bird of prey that has adapted to urban life as well as the kestrel is the owl. In North America it is the Common Screech Owl. In Europe it is the Tawny Owl (right).

Unlike kestrels, owls are *nocturnal* birds of prey. They sleep during the day and hunt at night. But like the kestrel, they eat small mammals. And both owls and kestrels use their ears in their search for prey.

The Short-eared Owl lives on the outskirts of town and hunts during the day.

Another bird that does very well in cities is the crow. Kestrels and crows can be seen in the sky chasing each other. In fact, crows may attack kestrels and eat their eggs or young. Kestrels sometimes nest in old crows' nests.

Living with People

People have not always appreciated the kestrel. Kestrels are predators, and owners of game farms throughout history have tried to kill kestrels and other birds of prey by the thousands. There used to be a "bounty" or a money reward for those who killed kestrels. This no longer exists.

The more recent threat to birds of prey is the use of *pesticides*. These chemicals are sprayed to kill insects, but they are not very accurate. The chemicals get sprayed on other animals and may be digested by kestrels. The pesticides have had serious effects on the birds. One of the more serious effects has been on the eggs. In some cases, the eggshells have become so thin that the eggs are accidentally broken when the adults sit on them. Bans on the deadly chemicals have increased steadily, and the numbers of healthy birds are on the rise.

For centuries the art of *falconry* has been a positive influence on the kestrel population. Birds of prey are bred and trained for hunting. Kestrels, because of their small size and unique ability to hover, are good birds for this.

Life in the Town

Kestrels have adapted well to life around people, and it is easy to find them soaring near highways. The slopes along the edges are filled with the field mice and small birds that kestrels like to eat. The animals in the field feed on smaller animals and plants. The kestrel is rarely eaten or attacked by other animals, as shown in this food chain.

Food Chain

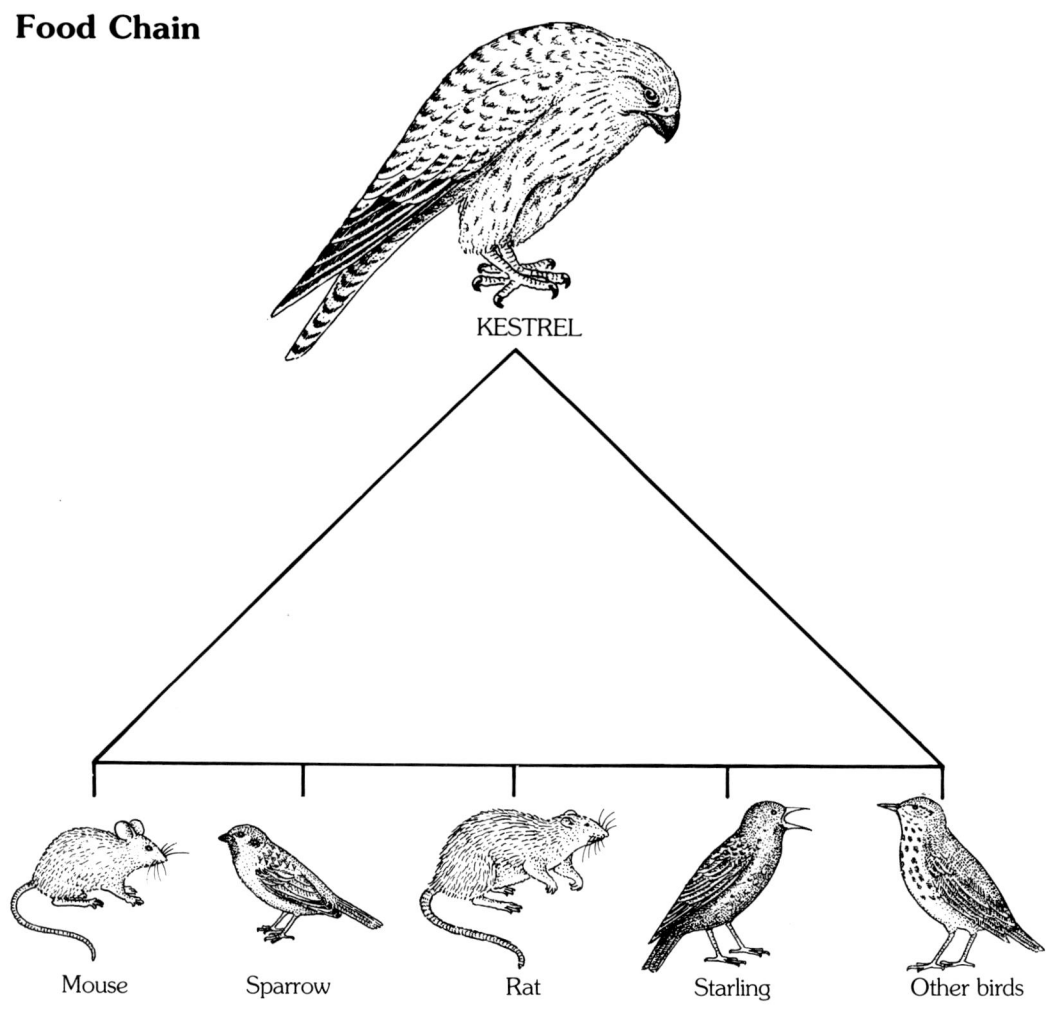

Kestrels love to nest on manmade structures like bell towers, cranes, office buildings, and even busy bridges with traffic roaring by. They cannot be disturbed in these places. And they can be found around the world, right in the heart of the cities, crying "kee, kee, kee!"

Index and New Words About Falcons

These new words about falcons appear in the text on the pages shown after each definition. Each new word first appears in the text in *italics*, just as it appears here.

air sacs .... large bags of air connected to the lungs and bone cavities which help to make a bird buoyant. **11**

brood patch .. a small patch of bare skin with a very rich blood supply on the underside of adult breeding birds. When a bird sits on its eggs to incubate them, it places its brood patch over the eggs so that heat is transferred to the eggs. **19**

carrion .... dead and rotting flesh. **15**

down (downy) .... fluffy feathers, as on young birds. **21**

falconry .... the art of training and flying birds of prey, mainly hawks and falcons, to hunt game for sport. **28**

fledge .... (of birds) to grow adult feathers before leaving the nest. **21, 22**

habitat .... the natural home of any group of animals and plants. **3, 5, 12**

incubate .... to keep eggs warm so that they will hatch. **19**

migrate .... (of animals) to move from one place to another at a regular time of the year, usually for breeding or overwintering. **22**

nocturnal ..... active at night. **26**

pesticides ..... chemicals used to poison insects or other pests. **28**

plumage .... the feathers covering a bird's body. **6, 7**

predator .... an animal that kills and eats other animals. **20, 24, 26, 28**

preening .... cleaning and oiling the feathers with the bill. **9**

prey .... animals that are hunted and killed by predators. **3, 8, 9, 10, 11, 13, 14, 15, 25, 26, 28**

roosting .... sleeping or resting. **13**

species .... a type of animal or plant that can interbreed successfully with others of its kind, but not with those of a different type. **3, 4, 25, 31**

stooping .... the steep diving flight of a hunting bird as it chases its prey. Usually birds like the kestrel will fly high above a flock of starlings and then "stoop" on their prey at great speed. **10**

talons .... sharp, curved claws belonging to birds of prey. **3, 15**

territory .... area that an animal defends against intruders. **16, 22**

urban .... living in a city or town. **5, 15, 19, 24, 25, 26**

Reading level analysis: SPACHE 2.8, FRY 4, FLESCH 85 (easy), RAYGOR 4, FOG 4, SMOG 3

Library of Congress Cataloging-in-Publication Data

Harrison, Virginia, 1966-
   The world of a falcon.

   (Where animals live)
   Adaptation of: The kestrel in the town.
   Includes index.
   Summary: Text and photographs present the characteristics and behavior of kestrels, including their feeding habits, home life, breeding techniques, and defense mechanisms.
   1. Kestrels--Juvenile literature. [1. Kestrels. 2. Falcons] I. Birkhead, Mike. Kestrel in the town. II. Oxford Scientific Films. III. Title. IV. Series.
QL696.F34H373   1988   598'.918   87-42611
ISBN 1-55532-333-2
ISBN 1-55532-308-1 (lib. bdg.)

North American edition first published in 1988 by Gareth Stevens Children's Books, 1555 North RiverCenter Drive, Suite 201, Milwaukee, Wisconsin 53212, USA. U.S. edition, this format, copyright © 1988 by Belitha Press Ltd. Text copyright © 1988 by Gareth Stevens, Inc. All rights reserved. No part of this book may be reproduced in any form or by any means without permission in writing from Gareth Stevens, Inc. First conceived, designed, and produced by Belitha Press Ltd., London, as **The Kestrel over the Town**, with an original text copyright by Oxford Scientific Films. Format copyright by Belitha Press Ltd. Series Editor: Mark J. Sachner. Art Director: Treld Bicknell. Design: Naomi Games. Cover Design: Gary Moseley. Line Drawings: Lorna Turpin. Scientific Consultant: Gwynne Vevers.

The publishers wish to thank the following for permission to reproduce copyrighted material: **Mike Birkhead** for back cover, title page, pp. 2, 3, 4, 6 both, 7 below, 8, 10, 11 both, 12 both, 13, 14 both, 15 both, 16, 17, 18 both, 19, 20, 21 both, 24 both, 26, 27 both, and 31; **Oxford Scientific Films Ltd.** for p. 5 (Wendy Neefus), p. 7 above (Leonard Lee Rue III), pp. 9, 23, and front cover (Patti Murray/Animals Animals), and p. 29 (D. G. Fox). For their help and advice, the author would like to thank the following: Caroline Aitzetmuller; Ashley Smith and Rosie the Kestrel from the Hawk Conservancy near Andover; Andy Village; and Ian Wyllie, David Quinn, Alastair MacEwen, and Jimmy Hull of the Oxford Museum. The publisher would like to thank the staff of the Havenwoods Forest Preserve, Wisconsin Department of Natural Resources, Milwaukee.

Printed in the United States of America
3 4 5 6 7 8 9 96 95 94 93 92 91

For a free color catalog describing Gareth Stevens' list of high-quality children's books, call 1-800-341-3569 (USA) or 1-800-461-9120 (Canada).

The world of a falcon /
J 598.918 H
31814850020766
Harrison, Virginia,
PORTAGE PUBLIC LIB   10112

J598.918   AUG 2 7 1992   Por.
    H
        Harrison, Virginia
            The world of a falcon

                Portage Public Library